FAST AND SLOW ANIMALS

BIRDS

BY BRENNA MALONEY

Children's Press
An imprint of Scholastic Inc.

A special thank-you to the team at the Cincinnati Zoo & Botanical Garden for their expert consultation.

Copyright © 2023 by Scholastic Inc.

All rights reserved. Published by Children's Press, an imprint of Scholastic Inc., *Publishers since 1920*. SCHOLASTIC, CHILDREN'S PRESS, and associated logos are trademarks and/or registered trademarks of Scholastic Inc.

The publisher does not have any control over and does not assume any responsibility for author or third-party websites or their content.

No part of this publication may be reproduced, stored in a retrieval system, or transmitted in any form or by any means, electronic, mechanical, photocopying, recording, or otherwise, without written permission of the publisher. For information regarding permission, write to Scholastic Inc., Attention: Permissions Department, 557 Broadway, New York, NY 10012.

Library of Congress Cataloging-in-Publication Data Available
Identifiers: LCCN 2022001770 (print)
ISBN 9781338836523 (library binding) | ISBN 9781338836530 (paperback)

10 9 8 7 6 5 4 3 2   23 24 25 26 27

Printed in China   62
First edition, 2023

Book design by Kay Petronio

Photos ©: 5 top: Tarpan/Dreamstime; 5 bottom: Alan Murphy/BIA/Minden Pictures; 6: Joe Riederer/Getty Images; 7: johnandersonphoto/Getty Images; 8–9: Joe Riederer/Getty Images; 12–13: Hiroya Minakuchi/Minden Pictures; 16–17: Mark Dumbleton/NIS/Minden Pictures; 20–21: Kevin Schafer/NPL/Minden Pictures; 24–25: Xi Zhinong/Nature Picture Library/Alamy Images; 26: Mike Warburton Photography/Getty Images; 30 bottom center: Tim Fitzharris/Minden Pictures.
All other photos © Shutterstock.

AMERICAN WOODCOCK

PEREGRINE FALCON

CONTENTS

Meet the Birds 4
#10 Slowest Bird: American Woodcock 6
American Woodcock Close-Up . . 8
#9: Chicken 10
#8: Gentoo Penguin 12
#7: Great Horned Owl 14
#6: Ostrich 16
#5: Anna's Hummingbird 18
#4: Grey-Headed Albatross 20
#3: Golden Eagle 22
#2: Saker Falcon 24
#1 Fastest Bird: Peregrine Falcon . . 26
Peregrine Falcon Close-Up 28
Birds Fast and Slow 30
Glossary 31
Index 32

MEET the BIRDS

Welcome to the world of birds! There are many types of birds, but all birds have some things in common. All birds have **backbones**, or spines. They all have wings and feathers. Instead of teeth, they have **beaks**. All birds hatch from eggs. Many birds build nests for their young. Birds are also **warm-blooded**. This means their body temperature does not change with their surroundings.

How Birds Move

All birds can move! But... how do they do it? Some birds walk or run on two legs, like ostriches. Some swim with flippers, like penguins. Most birds fly. Flying helps birds flee from danger. It also helps them hunt. Birds can move from cold places to warmer places. This is called **migration**. Birds move in many ways! Get ready to discover how 10 birds can travel, from the slowest to the fastest!

FACT Scientists who study birds are called **ornithologists** (or-nuh-THAH-luh-jists).

#10 Slowest Bird: AMERICAN WOODCOCK

The American woodcock is the world's slowest bird. Its top speed is 5 miles per hour (8 kph). To compare, a human walks at an average speed of 2.8 miles per hour (4.5 kph). This bird spends most of its time hidden in fields or on the forest floor. There, it pokes the ground with its beak, looking for worms to eat.

When it walks, the woodcock walks slowly. It stomps heavily with its front foot. This motion can make worms stir under the soil. That makes it easier for the bird to find them. Woodcocks are **nocturnal**. This means they are active at night. They fly low in the sky alone or in small **flocks**, or groups.

FACT: The American woodcock lives in North America.

AMERICAN WOODCOCK CLOSE-UP

FEATHERS
Light brown, black, and gray feathers help woodcocks blend in with the ground. This **camouflage** protects them from **predators**.

WINGS
Wings are broad and rounded.

Woodcocks probably do not drink water. They get enough water from the foods they eat.

FACT
A woodcock can open and close its beak to grasp **prey** while its beak is still in the ground.

#9 CHICKEN

FACT: Chickens can recognize the faces of up to 100 people.

Chickens *can* fly. Not very far and not very fast. But they can run up to 9 miles per hour (14.5 kph) in short bursts. They can also make sudden, short turns that keep them safe from predators. Chickens stay aware of their surroundings by bobbing their heads. They can't move their eyes, so they need to move their whole heads to change their view. By thrusting their head and eyes first, they can look for danger while their bodies catch up.

FACT Chickens live all over the world. They outnumber humans by almost 3 to 1.

Gentoo penguins may make as many as 450 dives a day to look for food.

FACT

#8 GENTOO PENGUIN

FACT: Gentoos live in and around Antarctica.

Gentoo penguins are thought to be the fastest-swimming penguins in the world. They have been estimated to reach speeds of up to 22 miles per hour (35.4 kph). For comparison, most penguins can swim about 4-7 miles per hour (6.4-11.3 kph). Gentoos are covered in feathers that block water. Water flows over their streamlined bodies. When they swim, they tuck up their **webbed** feet. Strong, paddle-shaped flippers pull them through the water.

#7 GREAT HORNED OWL

The great horned owl is the fastest owl. It is nocturnal and hunts at night. It can hear sounds from 900 feet (274.3 m) away. Like other owls, the great horned owl flies almost silently. Its soft feathers are arranged in a special way so that air flows over them quietly. Great horned owls swoop down to snatch prey with their **talons**. At top speed, they can fly up to 40 miles per hour (64.4 kph).

FACT
Great horned owls can be found throughout North and South America.

FACT Owls cannot move their eyes up, down, or side to side, so they must move their heads to see.

The ostrich can jog at 30 miles per hour (48.3 kph) for up to half an hour. **FACT**

FACT: Wild ostriches live in the dry, hot **savannas** and woodlands of Africa.

#6 OSTRICH

The fastest bird on land is the ostrich. It can sprint 45 miles per hour (72.4 kph). An ostrich can't fly, but it does have wings. The average **wingspan** of an ostrich is more than 6.5 feet (2 m).

Those wings help it run faster. Long legs give the ostrich its long stride. But its wings help it steer and balance as it speeds along.

#5 ANNA'S HUMMINGBIRD

The Anna's hummingbird is the fastest hummingbird in the world. It can reach speeds of 50 miles per hour (80.5 kph). It moves its wings in a figure-eight pattern. The Anna's hummingbird can **hover** over flowers to feed on **nectar**. When diving, the Anna's hummingbird covers the length of its own body 385 times per second. By comparison, the fastest jet can only cover 39 body lengths per second!

Hummingbirds are the only birds that can fly backward.
FACT

FACT Anna's hummingbirds can be found along the western coast of North America.

Grey-headed albatrosses live in Antarctica and across the islands of the Southern Ocean.

#4 GREY-HEADED ALBATROSS

FACT: The grey-headed albatross might fly more than 8,000 miles (12,874.8 km) in search of food.

The grey-headed albatross is one of the world's fastest horizontal flying birds. It can fly at speeds up to 79 miles per hour (127.1 kph). This bird spends nearly its entire life at sea. It can fly around the world in just a little over a month. Albatrosses save energy by rarely flapping their wings. Instead, they glide in the air. The average wingspan of a grey-headed albatross is 7 feet (2.1 m).

#3 GOLDEN EAGLE

The golden eagle is extremely fast in flight. This bird of prey can glide at speeds up to 80 miles per hour (128.7 kph). The golden eagle is also flexible. As they fly, golden eagles often hold their wings up in a slight V shape. They can soar effortlessly for hours. When they spot prey, they speed up to dive.

FACT

The golden eagle is North America's largest bird of prey and the national bird of Mexico.

FACT A golden eagle's wingspan is more than double the length of its body.

Sakers can be found from eastern Europe to central Asia. FACT

#2 SAKER FALCON

FACT The saker falcon is the national bird of Hungary and Mongolia.

The saker falcon is a patient hunter. This desert **raptor** catches prey close to the ground. It soars through the sky at up to 93 miles per hour (149.7 kph). When it spots small birds or rodents, it swoops down to snap up its food.

#1 Fastest Bird: PEREGRINE FALCON

The peregrine falcon is the world's fastest-diving bird. This record-setting falcon was once clocked diving at a speed of 186 miles per hour (299.3 kph). For comparison, most commercial helicopters fly at around 160 miles per hour (257.5 kph). This bird of prey hunts other birds by dropping down on them from a great height.

As a hunter, the peregrine falcon is unmatched. It circles its prey from above. Then it dives down, flapping its wings with great force. Suddenly, it folds its wings. Its body looks like a teardrop falling from the sky. When the falcon nears its prey, it catches it in its talons.

FACT Peregrine falcons are among the world's most common birds of prey.

BONES
Hollow bones make the falcon lighter while flying.

FACE
White face has a black tear stripe on each cheek.

BEAK
Sharp, hooked beak tears food.

KEEL
A large keel (breastbone) supports powerful flight muscles.

PEREGRINE FALCON CLOSE-UP

WINGS
Stiff, pointed wing feathers slice through the air during flight.

BODY
A streamlined body helps with diving.

FACT Peregrine falcons live on all continents except Antarctica.

TALONS
Strong, curved talons grip prey.

A peregrine's eyesight is eight times better than a human's eyesight. This bird can spot prey from more than 1 mile (1.6 km) away.

BIRDS FAST AND SLOW

Now you know birds can move in many ways. They run. They walk. They swim, and they soar. Some are slow, but many are not. The animals in this book are only a handful of the birds on Earth. There are more than 10,000 types of birds. Make it your mission to learn even more about these amazing animals and how fast they can go!

GLOSSARY

backbone (BAK-bohn) a set of connected bones that runs down the middle of the back; also called the spine

beak (beek) the horny, pointed jaw of a bird

camouflage (KAM-uh-flahzh) a disguise or natural coloring that allows animals to hide by making them look like their surroundings

flock (flahk) a group of animals of one kind that live, travel, or feed together, as in a *flock* of birds

hover (HUHV-ur) to remain in one place in the air

keel the breastbone of a bird

migration (my-GRAY-shun) movement from one area to another at a particular time of year

nectar (NEK-tur) a sweet liquid from flowers that bees gather and make into honey

nocturnal (nahk-TUR-nuhl) active at night

ornithologist (or-nuh-THAH-luh-jist) a scientist who studies birds

predator (PRED-uh-tur) an animal that lives by hunting other animals for food

prey (pray) an animal that is hunted by another animal for food

raptor (RAP-tur) a bird (such as an eagle or hawk) that kills and eats other animals for food

savanna (suh-VAN-uh) a flat, grassy plain with few or no trees

talon (TAL-uhn) a sharp claw of a bird such as an eagle, hawk, or falcon

warm-blooded (WORM bluhd-id) having a warm body temperature that does not change, even if the temperature of the surroundings is very hot or very cold

webbed (webd) having toes that are connected by a web or fold of skin

wingspan (WING-span) the distance between one end of a wing and the end of the other

INDEX

Page numbers in **bold** indicate images.

A
American woodcock, 6–9, **6–9**, **30**
 diet and hunting, 6–9
 habitat, 6–7
 movement and speed, 6–7
 physical traits, 8–9
 sleep, 7
Anna's hummingbird, **4**, 18–19, **18–19**, **30**

B
birds
 common traits, 4
 number of, 30
 ways of moving, 5, 30

C
chicken, **4**, 10–11, **10–11**, **30**

G
Gentoo penguin, **5**, 12–13, **12–13**, **30**
golden eagle, **4**, 22–23, **22–23**
great horned owl, **4**, 14–15, **14–15**
grey-headed albatross, 20–21, **20–21**

O
ostrich, **5**, 16–17, **16–17**

P
peregrine falcon, **5**, 26–29, **26–29**, **30**
 diet and hunting, 26–27
 habitat, 29
 movement and speed, 26
 physical traits, 28–29

S
saker falcon, 24–25, **24–25**

ABOUT THE AUTHOR

Brenna Maloney is the author of more than a dozen books. She lives and works in Washington, DC, with her husband and two sons. She is about as slow as an American woodcock but wishes she were as fast as a peregrine falcon.